I0644148

The World's Oceans

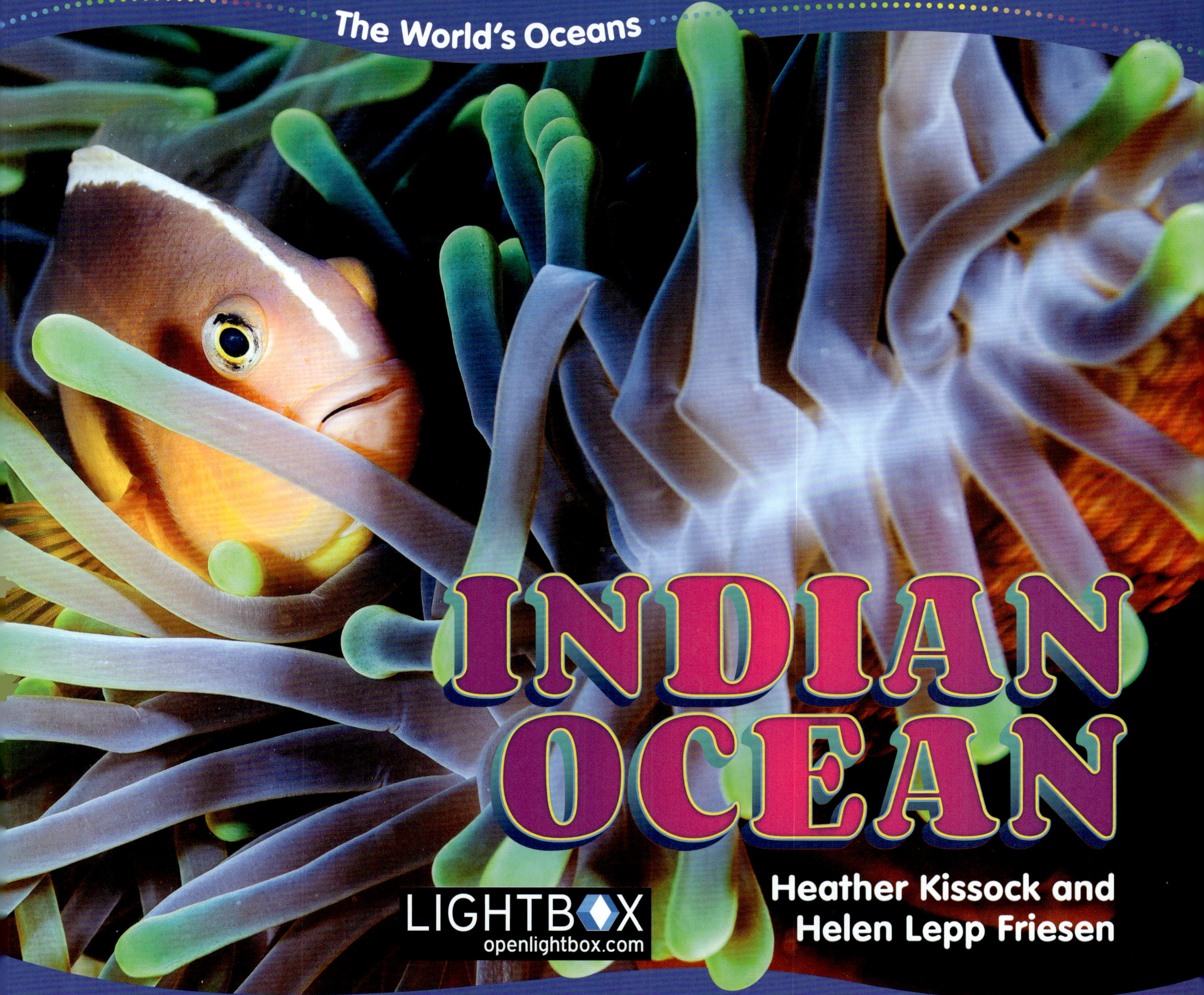

INDIAN OCEAN

Heather Kissock and
Helen Lepp Friesen

LIGHTBOX
openlightbox.com

LIGHTBOX

Go to **www.openlightbox.com** and enter this book's unique code.

ACCESS CODE

LBXR3575

Lightbox is an all-inclusive digital solution for the teaching and learning of curriculum topics in an original, groundbreaking way. Lightbox is based on National Curriculum Standards.

OPTIMIZED FOR

- ✓ TABLETS
- ✓ WHITEBOARDS
- ✓ COMPUTERS
- ✓ AND MUCH MORE!

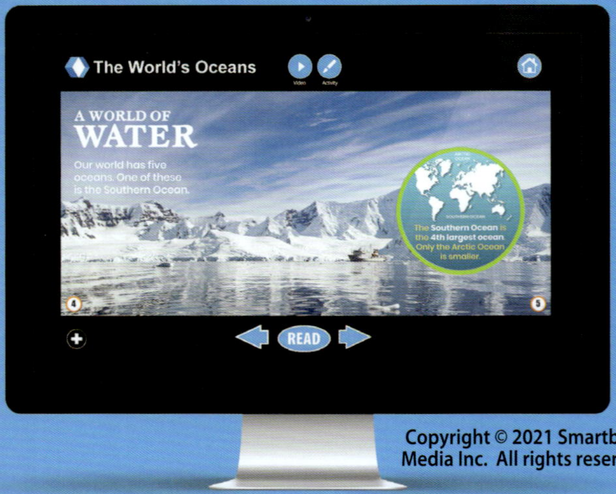

Copyright © 2021 Smartbook Media Inc. All rights reserved.

STANDARD FEATURES OF LIGHTBOX

AUDIO High-quality narration using text-to-speech system

VIDEOS Embedded high-definition video clips

ACTIVITIES Printable PDFs that can be emailed and graded

WEBLINKS Curated links to external, child-safe resources

SLIDESHOWS Pictorial overviews of key concepts

INTERACTIVE MAPS Interactive maps and aerial satellite imagery

QUIZZES Ten multiple choice questions that are automatically graded and emailed for teacher assessment

KEY WORDS Matching key concepts to their definitions

VIDEOS

WEBLINKS

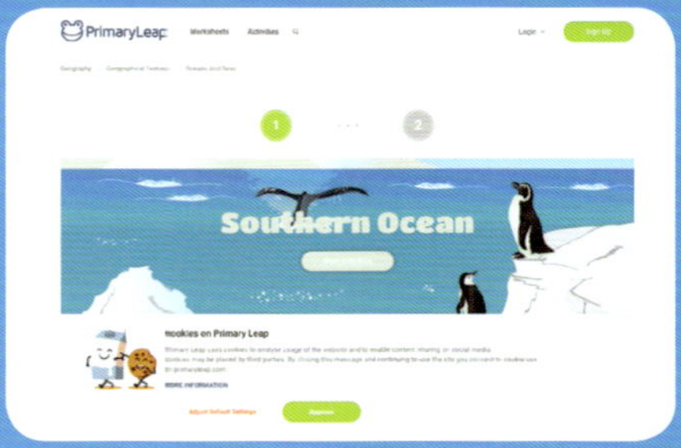

SLIDESHOWS

QUIZZES

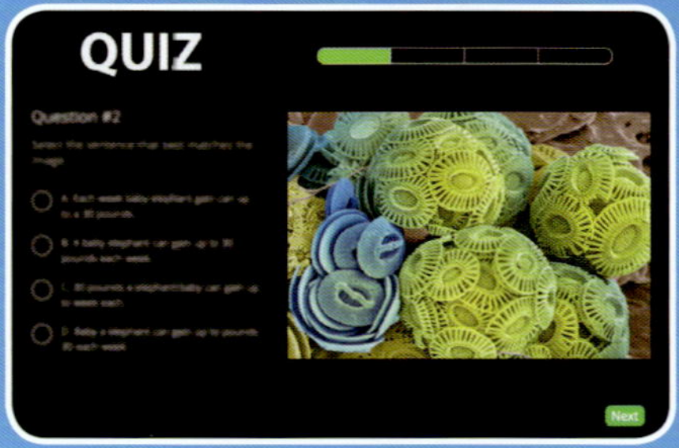

The World's Oceans

INDIAN OCEAN

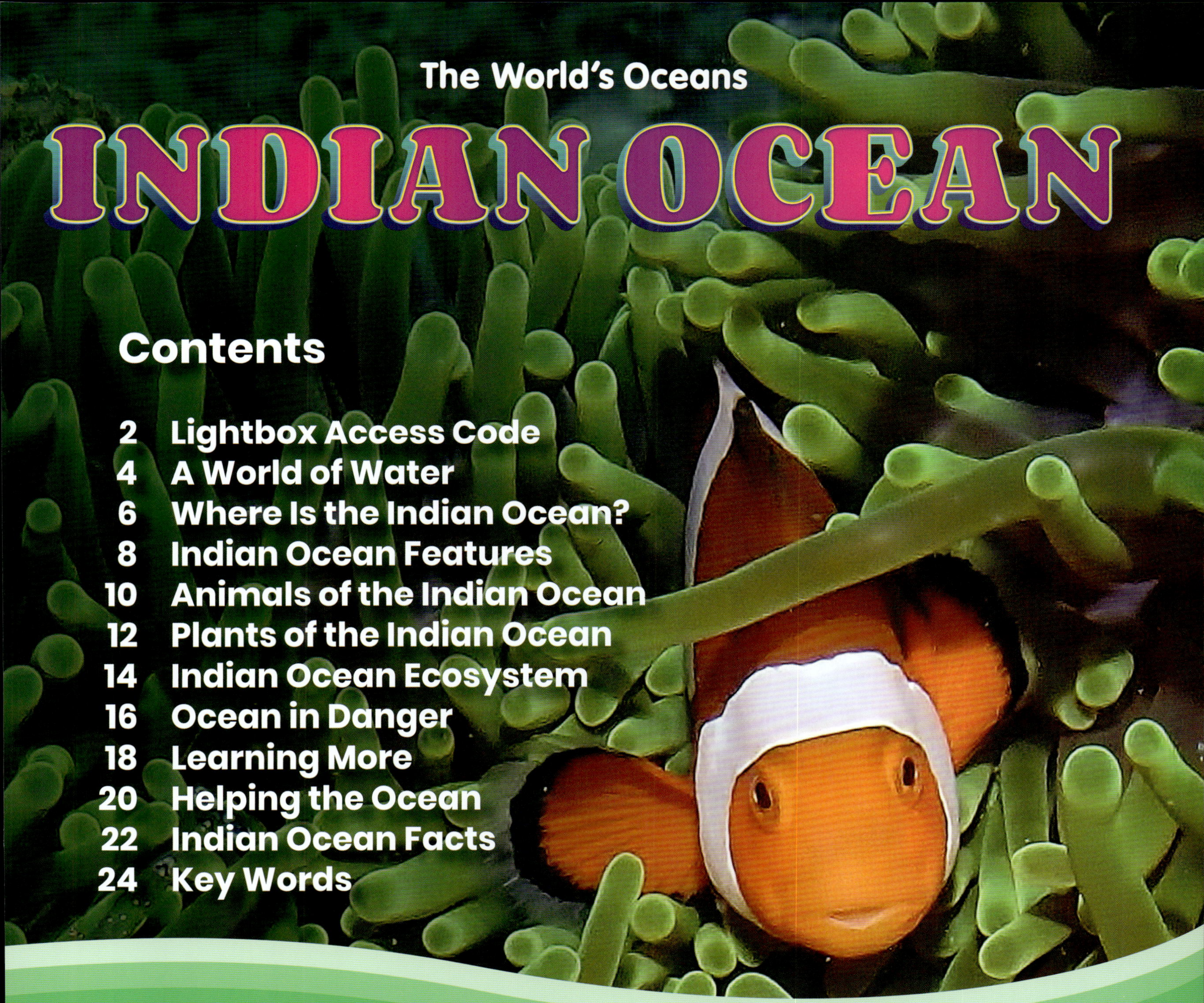

Contents

2 Lightbox Access Code
4 A World of Water
6 Where Is the Indian Ocean?
8 Indian Ocean Features
10 Animals of the Indian Ocean
12 Plants of the Indian Ocean
14 Indian Ocean Ecosystem
16 Ocean in Danger
18 Learning More
20 Helping the Ocean
22 Indian Ocean Facts
24 Key Words

A WORLD OF WATER

Our world has five oceans. The Indian Ocean is the world's warmest ocean.

The **Indian Ocean** can be as **warm** as **82° Fahrenheit (27.7° Celsius)**. That is **beach weather!**

WHERE IS THE INDIAN OCEAN?

The Indian Ocean touches the continents of Africa, Asia, and Australia. It links to three other oceans.

THE WORLD'S OCEANS

INDIAN OCEAN FEATURES

The Indian Ocean has several underwater volcanoes. Some even rise above the water.

The **Indian Ocean**'s tallest underwater volcano has a **height** of about **16,400 feet** (5,000 meters).

ANIMALS OF THE INDIAN OCEAN

The Indian Ocean is home to a variety of animals.

Sea Sponge

Hawksbill Turtle

Tiger Shark

Giant Clam

Clownfish

PLANTS OF THE INDIAN OCEAN

Many unique plants grow in the Indian Ocean. They provide food for the animals that live there.

Coralline Algae

Phytoplankton

Sea Lettuce

Eelgrass

INDIAN OCEAN ECOSYSTEM

The plants and animals that live in the Indian Ocean need each other to survive. Together, they form an ecosystem.

INDIAN OCEAN FOOD CHAIN

Phytoplankton grows in shallow waters.

Phytoplankton is food for sea sponges.

Sea sponges are a hawksbill turtle's main food.

Hawksbill turtles are a source of food for tiger sharks.

OCEAN IN DANGER

More people are moving to be near the Indian Ocean.

Ocean animals are losing their homes and the clean water they need to survive.

LEARNING MORE

People are working to help the animals in the Indian Ocean.

They are setting aside areas where the animals will be safe.

HELPING THE OCEAN

You can help the Indian Ocean and its plants and animals.

Work to keep the waterways close to you clean.

All of Earth's waters flow into the global ocean.

INDIAN OCEAN FACTS

This page provides more detail about the interesting facts found in the book. Simply lock at the corresponding page number to match the fact.

Pages 4–5

The Indian Ocean is the third-largest of the world's oceans. It covers an area of approximately 28,360,000 square miles (73,450,000 square kilometers). The Indian Ocean is bordered by land to the north. Without a connection to the cold Arctic Ocean, the northern regions of the Indian Ocean are warm.

Pages 6–7

The Asian countries of India, Pakistan, Iran, and Bangladesh border the Indian Ocean to the north. Indonesia, Australia, and the Malay Peninsula lie to the east. In the west, the Indian Ocean is bordered by Africa and the Arabian Peninsula. The Southern Ocean borders the Indian Ocean in the south. The Pacific Ocean lies to its east and the Atlantic to the west.

Pages 8–9

Seamounts are inactive mountain volcanoes under the ocean's surface. The Muirfield Seamount is the Indian Ocean's tallest seamount. Its peak comes within 60 feet (18 m) of the water surface. Other seamounts have one or two flat tops that jut above the ocean's surface. Volcanoes formed most of the islands in the Indian Ocean.

Pages 10–11

The animals of the Indian Ocean depend on its warm water and climate to survive. Clownfish live in the Indian Ocean's tropical coral reefs, where they hide in the tentacles of anemones. Hawksbill turtles can also be found in shallow reef waters. They find much of their food here.

Pages 12–13

Hundreds of species of plants make their home in the Indian Ocean. Many animals rely on these plants for food. Phytoplankton is the main food source for most herbivores in the Indian Ocean. Eelgrass is found in shallow waters. Besides being a food source, it also serves as a home for crabs and small fish.

Pages 14–15

An ecosystem is a group of living beings. These organisms interact with each other and their surroundings. Within this system there are producers and consumers. Plants are producers because they use the Sun's energy. Animals are consumers. They eat what other organisms produce. This cycle of producing and consuming is called a food chain.

Pages 16–17

By 2040, the urban population in Eastern Africa is expected to be five times higher than in 2010. The resulting coastal development for roads, buildings, and other structures has the potential to damage or demolish mangroves, wetlands, coral reefs, and other wildlife habitats. Increasing amounts of domestic sewage will pollute coastal waters as well.

Pages 18–19

Governments and other groups are making an effort to protect the animals and plants of the Indian Ocean. The Indian Ocean is now home to several Marine Protected Areas (MPAs). These government-created areas protect parts of the ocean from human activity so that marine plants, animals, and ecosystems can flourish.

Pages 20–21

Approximately 97 percent of Earth's water forms one major body called the global ocean. Scientists divide the global ocean into five basins. These are known as the Atlantic Ocean, Pacific Ocean, Indian Ocean, Arctic Ocean, and Southern Ocean. Water moves freely through these oceans, as do many of the animals that live in them.

KEY WORDS

Research has shown that as much as 65 percent of all written material published in English is made up of 300 words. These 300 words cannot be taught using pictures or learned by sounding them out. They must be recognized by sight. This book contains 54 common sight words to help young readers improve their reading fluency and comprehension. This book also teaches young readers several important content words, such as proper nouns. These words are paired with pictures to aid in learning and improve understanding.

Page	Sight Words First Appearance
4	a, has, is, of, our, the, water, world
5	as, be, can, that
6	and, it, other, three, to, where
8	above, even, some
9	about, feet
10	animals, home
12	food, for, grow, in, live, many, plants, there, they
13	sea
14	an, each, need, together
15	are
16	more, near, people
17	their
18	help
19	will
20	its, you
21	all, close, Earth, into, keep, work

Page	Content Words First Appearance
4	Indian Ocean, oceans
5	weather
6	Africa, Asia, Australia, continents
7	Antarctica, Arctic Ocean, Atlantic Ocean, Europe, North America, Pacific Ocean, South America, Southern Ocean
8	features, volcanoes
9	height, meters
10	variety
11	clownfish, giant clam, hawksbill turtle, sea sponge, tiger shark
13	coralline algae, eelgrass, phytoplankton, sea lettuce
14	ecosystem
15	food chain, source
16	danger
19	areas
21	waterways

Published by Smartbook Media Inc.
14 Penn Plaza 9th Floor New York, NY 10122
Website: www.openlightbox.com

Copyright ©2021 Smartbook Media Inc.
All rights reserved. No part of this publication may be reproduced, stored in a retrieval system, or transmitted in any form or by any means, electronic, mechanical, photocopying, recording, or otherwise, without the prior written permission of the publisher.

Library of Congress Cataloging-in-Publication Data
Names: Kissock, Heather, author. | Friesen, Helen Lepp, 1961- author.
Title: Indian Ocean / Heather Kissock and Helen Lepp Friesen.
Description: New York, NY : Smartbook Media, Inc., 2021. | Series: The world's oceans | Audience: Ages 6-8 | Audience: Grades 2-3 |
Identifiers: LCCN 2020018628 (print) | LCCN 2020018629 (ebook) | ISBN 9781510555204 (library binding) | ISBN 9781510555211
Subjects: LCSH: Indian Ocean--Juvenile literature. | Oceanography--Indian Ocean--Juvenile literature. | Marine ecology--Indian Ocean--Juvenile literature.
Classification: LCC GC721 .K57 2021 (print) | LCC GC721 (ebook) | DDC 551.46/15--dc23
LC record available at https://lccn.loc.gov/2020018628
LC ebook record available at https://lccn.loc.gov/2020018629

052020
110819

Printed in Guangzhou, China
1 2 3 4 5 6 7 8 9 0 24 23 22 21 20

Editor: Heather Kissock Designer: Jean Rodriguez

Every reasonable effort has been made to trace ownership and to obtain permission to reprint copyright material. The publisher would be pleased to have any errors or omissions brought to its attention so that they may be corrected in subsequent printings. The publisher acknowledges Alamy, Shutterstock, Getty Images, and iStock as its primary image suppliers for this title.